THE

LIGHTNING CALCULATOR:

A

New, Readable, and Valuable Book,

CONTAINING

THREE NEW PROCESSES OF ADDITION, FOUR NEW FORMS OF MULTIPLICATION, RAPID PROCESSES OF SQUARING AND CUBING, SUBTRACTION AND DIVISION, HIS IMPROVED FORM OF INTEREST, AND VALUABLE INFORMATION IN BOOK-KEEPING;

TOGETHER WITH

A HISTORY OF HIS REMARKABLE LIFE, HIS WONDERFUL DISCOVERIES IN NUMBERS, HIS AMUSING AND INSTRUCTIVE PARLOR FEATS, ETC., WITH HIS AUTOGRAPH.

JUST ISSUED BY

PROFESSOR HUTCHINGS.

New York:
CLARRY & REILLEY, PRINTERS AND ENGRAVERS,
Nos. 12 & 14 Spruce Street.
1867.

A BRIEF HISTORY OF MY EVENTFUL LIFE.

I was born in the City of New York, at the corner of Eldridge and Hester streets, on Saturday, January 7th, 1832. My father was a Long Island man and my mother a Connecticut woman. My father was for many years a wholesale and retail merchant in New York City. I will say here, in passing, that at Hubbs & Clark's Academy, where I graduated, I showed some little ability—more than ordinary—in figures; but I will also say that I attribute my present skill in figures to careful research and untiring perseverance on my part, and I earnestly commend my readers to perfect diligence and thorough practice as the surest road to success. I was for some years a book-keeper in my father's counting-room, but it was not until some three years since that my attention was particularly called to figures. I have in my present work embodied the various short processes that I have gathered together in the last three years; and as far as short processes and rapidity of execution go, I do most heartily commend it to the public, as by far the best work ever published on this subject.

WM. S. HUTCHINGS.

A GUIDE

TO

Rapid and Accurate Computation,

BY

PROFESSOR HUTCHINGS,

THE MATHEMATICAL PHENOMENON AND LIGHTNING CALCULATOR.

N.B.—It is particularly understood that the possessor of this secret has pledged his or her honor not to divulge the contents of the following lines to any other person or persons, but rather induce them to obtain it from the discoverer, Professor Hutchings, for the same price which has been paid by the owner of this copy.

PREFACE.

Few things are impossible to the human mind. The spark of intellect with which the Creator has endowed us all may be fanned into a large flame by cultivation, practical experience, and application. The act of computation, or addition, more so than any other, is accessible to every one possessed of a vigorous and active mind, and every one may be enabled to cast up figures or add sums together with lightning velocity by observing a few rules established by the Professor, after a close study of the figures, and some not overhard practice. There is a way of doing a thing well and to the purpose, and this applies more particularly to computation. This way will be indicated distinctly in the following lines, and all that will astonish the reader will be the fact that the simplicity and practicability of these few rules has not found its way long before this into our schools and offices. It will not be expected that the student be ignorant of his Multiplication Table, or that "two and two are four"—for that our children go to school. But as a hint, by way of introduction, it may be remarked, that in adding up, for instance, the numbers 5, 6 and 7, it is quite as easy to say 5, 11, 18, as it is to proceed in the old-fashioned way: 5 and 6 are 11 and 7 are 18, and considerably quicker and more to the purpose as all will acknowledge. Here are but three simple rules, contained in a few words, which put the reader at once in possession of all the means which will enable him, with some practice, to acquire the same expeditious way of computing numbers most accurately, of which he may have seen an example at Barnum's American Museum.

RULE I.—Commence at the bottom, run to the top and down again. In no case set down the carrying figure until you arrive at the bottom, and then set down both results. Make use of the hint thrown out in the Preface; that is, if you have to add the following numbers: 2, 9, 7, 4, 3, 8, 1, 6, do it in this manner: 2, 11, 18, 22, 25, 33, 34, 40.

RULE II.—Commence at the top, run down to the bottom, up again and down again, and so on until the computation is completed.

RULE III.—In computing a column composed, for instance, of these figures, 9, 8, 4, 6, 5, 6, 4, 3, connect two figures in your mind and proceed thus: 17, 27, 38, 45—9 and 8 being 17, 4 and 6 being 10, 5 and 6, 11—and so on, treating these single results in the way indicated in Rule I.

These rules will aid you very materially in gaining rapidity. You will now commence your study in good earnest.

Supposing you to be very familiar with the results of every two figures up to 20, you will then sit down quietly and compute on your slate every two figures between 20 and 30, thus: 20 and 1, 20 and 2, 20 and 3, up to 20 and 9; then again, 21 and 1, 21 and 2, 21 and 3, up to 21 and 9, until you arrive at 29 and 9, impressing, as you proceed, each count upon your mind until you are perfect. You will then go through the same process with the figures between 30 and 40, which done, set down six columns of the figure two, with 6 in each column; the same of the figures 3, 4, 5, 6, 7, 8, and 9, and compute each until you are rapid and accurate in every one.

You will now practice the three following examples twelve times, according to Rule I., six times as indicated in Rule II., and six times according to Rule III., gaining rapidly at each successive trial:

879	567	745
542	768	484
363	699	363
685	454	576
456	565	662
947	646	291

Having done this, compute on your slate from 40 to 50 as before, and again from 50 to 60; after which practice the three following examples, according to Rule I., ten times; Rule II., twelve times; and Rule III.

ten times; and you will discover more ease and rapidity in your computation than you have ever before enjoyed:

967	787	478
967	787	874
967	999	787
976	579	666
858	468	599
989	378	489
887	796	379

Resume now your practice with the figures between 60 and 70, 70 to 80, 80 to 90, 90 to 100. Having thus gone over the whole addition table, you will be familiar with every combination of figures in it. Then practise the three following examples as follows: Rule I., twenty times; Rule II., fifteen times; Rule III., twenty times; and you will be astonished at your rapidity:

2222	8767	998769
3333	4579	876879
4444	9969	787987
5555	8888	859786
6666	7777	674875
7777	6666	467987
8888	5432	896754
9999	7654	979886

MULTIPLICATION.

32
23
——
736

PROCESS—TWO FIGURES.—Units into units will give first figure; units into tens and tens into units will give second figure; tens into tens will give third figure.

323
232
——
74,936

PROCESS—THREE FIGURES—Units into units will give first figure; units into tens and tens into units will give second figure; units into hundreds and hundreds into units, and tens into tens added, and multiplied with the carrying figure, will give third figure; tens into hundreds and hundreds into tens will give fourth figure; hundreds into hundreds will give fifth figure.

Always observe great care in carrying.

SECRET FOR BOOK-KEEPERS.

In the practice of book-keepers to transfer whole columns of sums from the Day Book to the Ledger, etc., etc., it will frequently be found, that, after adding up the columns in the two books, the sum-total differs. Now, when such is the case, and the book-keeper wishes to ascertain whether the mistake has been made in the computation or in the extension of the single counts, he has but to subtract the lesser sum total from the larger, and divide by 9. If there be a remainder, the mistake is made in the computation; if not, it lies in the extension of the single counts, which then must be compared. For instance:

376	376	376	376
482	482	482	482
190	100	190	190
397	397	397	397
454	454	454	454
132	123	132	132
654	654	654	654
323	323	323	323
875	875	875	875
950	905	950	950
4833	4698	4833	4843
4698			4833
9)135(15			9)10(1
9			9
45			1 Remainder,
45 No remainder,			which proves the computation incorrect.

which proves the computation correct.

Rule of Addition for Two Columns.

32
43
56 24 and 6 = 30 and 50 = 80 and 3 = 83 and 40 = 123 and
24 2 = 125 and 30 = 155.

155

Rule of Addition for Three Columns.

223
320
234 20 and 34 = 54 and 5 = 554 and 2 = 754 and 20 = 774
520 and 3 = 1074 and 23 = 1097 and 2 = 1297.

1297

These methods I consider quite expeditious and extremely simple, and consequently very valuable to all minds.

Rule of Multiplication for any Two Figures in the Multiplier, and any Number of Figures in the Multiplicand.

Example: 324 by 23.
972
———
7452

First, multiply your multiplicand by the right-hand figure of multiplier 23, and extend one place right; then multiply by left-hand figure of multiplier, and take in partial product as you go through.

When the Multiplicand has Four Figures and the Multiplier Three.

2243
0222
———
497946

Simply proceed as directed in the previous cross multiplication, only let the cypher stand in the thousand's place, as in above example. This is very useful.

To Multiply any Number of Figures by 50, 25, *or* 12½.

For 50, add two cyphers, and divide by two; for 25, by 4; for 12½ by 8.

Examples: 222 by 50. 2)22200
———
11100

222 by 25. 4)22200
———
5550

222 by 12½. 8)22200
———
2775

This is sometimes very convenient.

Squaring 24 = 576. *Units into units, double of units by tens, tens into tens; always carry.*

To Square any Number of Figures ending with 5.

Simply multiply the left-hand figure or figures by the next higher, and annex 25, which is the square of 5:

45 = 4 by 5 = 2025.
65 = 6 by 7 = 4225.

325 = 32
33
———
105625

To Square any Figure of Figures ending with ½.

Multiply the left-hand figure or figures by the next higher, and annex ¼.

Square of 23½ = 552¼

23
24
552¼

Cubing.—First, square your figures as directed above; then multiply that by the two-figure principle, as you have been previously shown:

Cube of 22 = 484
968
10648

Two-figure rule. Multiply your multiplicand by right-hand figure of multiplier; extend one place right; then multiply by left-hand figure, and add in partial products.

EGG QUESTION.—A man has 3 boys. To one, he gives 10 eggs; to another, 30 eggs; to another, 50 eggs. Each is to sell his eggs at the same price, and return the same money.

Solution.

EGGS.	EGGS.	EGGS.
10	30	50
7	28	49
9	6	3
CTS.	CTS.	CTS.
1	4	7
9	6	3
10 cts.	10 cts.	10 cts.

The first sells his at 7 for 1 cent.

Next sells at same price 4 cents' worth.

Next, 49 at 7 for a cent.

Secondly, they all sell their eggs at 3 each.

WONDERFUL AND INSTRUCTIVE FEATS.

Write the Answer to a Sum before the Sum is put down at all.

324 685 964 314 035 —— 2322	Let any person write one row of three figures. Then, to write answer, take 2 from right-hand top figure, and bring down balance; bring down the other two figures, and always place 2 at left hand, as in example; then let the other party write two rows more, and you will make second and third rows come to 9.
3464 8653 9567 8643 1846 0432 1356 —— 33461	You may let a person write a square of sixteen figures, four wide and four deep; then you will write four wide and three deep; and instantaneously write the answer by taking 3 from top right-hand figure; bring down balance with other figures, and placing the 3 at left hand.

You may let a person write down any three figures or more, add them together and place the sum of them underneath, and subtract it from first figures; he will then scratch out any one figure of the answer, and give you the remaining figures. You will then add together the figures given you, and subtract their sum from 9 or 18, or the extended multiple of 9, as in examples:

225	2
9	1
—	—
216	3 from 9 = 6 scratched out.
987	9
24	6
—	—
963	15 from 18 = 3 scratched out.

To Produce the Answer all One Figure.

```
 12345679
       21
---------
 12345679
 24691358
---------
259259259
        3
---------
777777777
```

Multiply the figure called for by 9; then multiply the above figures by that multiplier or its factors.

You may let a person write down any number of figures for a multiplicand; then have the party multiply it by any two figures or more, which, when multiplied together will produce a multiple of 9, as in example; then let the party rub out all the upper work, except the last result; then let the party rub out any one figure of the last result, and you, by seeing and adding the remaining figures, and subtracting their sum from the next multiple of 9, may tell the figure rubbed out.

```
 3423
    6
-----
20538
    3
-----
61614
```

```
                   P. F. J.
                   2, 4, 2.
          Mul. by 2
                  —
                  4
Add 5..........   5
                  —
                  9
Multiply by 5....  5
                  —
                 45
Add finger......  4
                  —
                 49
Multiply by 10...10
                 ——
                490
Add joint......   2
                 ——
                492
Subtract........250
                 ——
                242
```

Ring Feat.—Select from a party a number of persons, say four or more; number them off 1, 2, 3, 4, and so on; the head of the class at your left hand; allowing each 10 fingers, including thumbs; let a person follow the form, give the result, subtract 250, and tell, who has the ring.

CERTIFICATE OF EDWARD EVERETT, AND OPINIONS OF THE PRESS.

Mr. Hutchings has exhibited to me a specimen of his skill in Arithmetical Calculation, which is very remarkable.

EDWARD EVERETT.

BOSTON, 22d July, 1862.

[*From the New York Tribune, April* 19.]

A PHENOMENON.—This is not a gift, but a scientific process, which he can impart to pupils. It will be of immense advantage in trade, commerce, and science, and revolutionize the tedious mode of addition throughout the world.

[*From the Scientific American.*]

We have examined his processes, and are satisfied that almost any of our intelligent accountants, who are pretty quick at figures, might learn to calculate with nearly the same rapidity.

www.ingramcontent.com/pod-product-compliance
Lightning Source LLC
LaVergne TN
LVHW010833120826
845149LV00016B/1366

* 9 7 8 1 4 1 8 1 8 9 5 0 1 *